AF345654

Folding the Future: Machine Learning Revolutionizes Protein Structure Prediction

Mariya

Copyright © 2024 byMariya

All rights reserved. No part of this book may be
reproduced in any manner
whatso-ever without written permission except
in the case of brief quotations
embodied in critical articles and reviews.

First Printing, 2024

Table of Contents

Chapter 1 Introduction

Machine learning has been widely used in chemistry and bioinformatics, to leverage the patterns learned from large data sets. Compared to the computational methods which use the empirically fitted parameters, machine learning techniques are more robust. Beneficial from the development of the Graphic Processing Unit, larger datasets and more complex model architectures are being explored, and breakthroughs are coming out quickly.

A successful example is the protein folding solver: AlphaFold by DeepMind. In 2020 CASP14, AlphaFold2 achieved 87.0 GDT (Global Distance Test) in free modeling section which was seen as a big breakthrough in protein structure prediction, while the best scores in the 2006 - 2016 competitions were about 30 to 40 GDT in the same section. AlphaFold used deep learning to learn from the protein sequences and structures, together with biophysical information to predict how the proteins fold (Hutson 2019; Callaway 2020).

Besides the progress in accuracy, once trained the machine learning models can predict a new structure from the learned parameters faster compared to many *de novo* structure prediction methods, which is a big advantage when applied to a huge screening space.

In this book, I will present two projects where we have applied machine learning techniques to proteins. In the first project where we predicted the protein-protein interactions we started from protein sequences in *S. cerevisiae* and human, and applied pretrained language models to produce embeddings where the models have learned to extract the features from large protein sequence datasets. We also explored the performance of extracting features from the protein interaction graph. These features were then fed into different classification models to fulfill the interaction prediction task.

In the second project where we predicted the protein loop structures which are the most flexible parts in proteins and the most difficult parts in the protein structure prediction task, our training loops included all the available loops, in contrast to the homology modeling where loop templates were searched only within the homologs. Although AlphaFold achieved great performance in protein folding, loops were not taken care of specifically. We would like to take advantage of machine learning on the loop modeling task. The data source we used is the protein sequences and we have easy access to large datasets of them.

To be successful in applying machine learning techniques, there are three key factors: data, feature engineering, and model selection. To extract the features from protein sequences, we focused on the language models. As sentences are from combinations of words in the human language, a protein sequence consists of amino acids with a certain sequence. Different combinations of words make different sentences have different meanings, and different combinations of amino acids serve as the first structure level of proteins which is the foundation of protein 3-dimensional structures, properties and functions. It is natural to think that the models which are useful extracting information from human language can help to extract information from protein sequences as well.

Human languages consist of sentences, which have two features: the building blocks and the sequence. The building blocks are the words, and the group of words is called vocabulary for a specific language. The 'vocabulary' for protein sequences contains the primary protein amino acids. Different amino acids have different physical properties such as hydrophilicity, dissociation constant (Barrett 2012). It has also been shown that some mutations of amino acids caused inherited diseases because of the change of protein functions (Krawczak et al. 2000).

From these we can see that the distribution and the sequence of the amino acids in a protein contain the important information which can affect the protein structures and functions.

A protein sequence can be transformed into a one-dimensional array with each amino acid represented by a unique integer, or it can be transformed into a matrix with one-hot encoding. However direct use of such simple encoding with classification models usually fails to have a nice performance. One strategy is to use sequence profile which is the position-specific scoring matrix (PSSM) produced by multiple sequence alignment (MSA) such as its application in the protein secondary structure prediction (Jones 1999). Evolutional information was captured during the MSA process besides the information from a single sequence alone. However, the alignment process is needed every time preparing the sequence profiles for new samples, which is very time consuming. Other descriptors generated from biological and physical properties can be added as features as well, such as amino acid composition, hydrophobicity, volume, polarity, etc. (Du et al. 2017). In our research we didn't consider such descriptors for two reasons. The first is that it is tedious to identify the most suitable descriptors for our goal out of many possible choices; Secondly and the most important, many projects have shown the successful predictions of various protein functions (Kulmanov and Hoehndorf 2020) and properties such as solubility (Magnan, Randall, and Baldi 2009) from sequence-based methods. Thus, we decided not to include the property descriptors as features but to use only protein sequences.

Another strategy is to learn the protein representations by training a complex model on a large protein sequence dataset. Recurrent Neural Network (RNN) has been widely used in the language problems for its ability to learn the patterns from sequential data. Many deep RNN models have been trained on large sets of protein sequences in order to generate informatic representations for protein sequences such as ESM (Rives et al. 2021) and TAPE (R. Rao et al.

2019). Such methods transferred the self-supervised RNN models which were successful on learning representations for natural language to protein sequences and compared the performance of such learned protein representations on the protein-related downstream tasks. In the downstream tasks as secondary structure prediction, the performance of ESM and TAPE embeddings generally exceed the simple sequence encoding methods and is comparable to the MSA sequence profile method (R. Rao et al. 2019; Rives et al. 2021). We researched on how the performance of such language embeddings on the interaction prediction task is.

Besides the novel sequence representations from language models, we also imported the use of the protein interaction graph. It is assumed that if two proteins A and B share several interacting proteins, the interacting interfaces of the two proteins may be similar. If we could find another protein C that interacts with A, we then suppose B interacts with C with high chance as well. This idea is called L3 principle that the path between the two possible interacting proteins in the interaction graph has the length of three, and is different from the usual link prediction tasks in social network where two nodes sharing many neighbors are more likely to link (Kovács et al. 2019). We combined the language embeddings with such graph learning method to generate another feature for our task.

In the second project where we predicted the loop structures, we built and trained our own language models on the protein sequences. We used the sequences within and around the loops as inputs and predicted the loop inter-residue distances. Since it is a task starting and ending with sequential data, we chose the encoder-decoder model or seq2seq language model (Sutskever, Vinyals, and Le 2014) which was successful in language translation. The distances were represented by discrete bins with 0.1 difference, thus the decoder network actually did a classification task. To fill up the loop gap in an actual protein structure file, a full-atom loop build-up process was needed from the predicted inter-residue distances. We applied two different programs which did full-atom reconstruction and minimized the energy for the final loop conformations from the predicted distances. Such incorporation of machine learning predictions into physics-based methods helps to increase the accuracy and accelerate the loop build-up stage.

Chapter 2 Protein-Protein Interaction Prediction with Deep Learning Methods

This chapter is a draft paper for the research project about protein-protein interaction prediction with machine learning. We used the embeddings from pretrained language models on protein sequence database as the protein features in our methods, tested our methods on the *S. cerevisiae* dataset and the human dataset, and compared our results to other state-of-art methods.

2.1 Introduction

Protein-protein interactions (PPIs) are essential in many physiological processes, such as immune response and DNA synthesis. The detection and analysis of protein-protein interactions and the protein interaction networks have been successfully applied to numerous biomedical fields, such as protein function predictions (Hishigaki et al. 2001; Vazquez et al. 2003), gene-disease associations assignment (J. Chen, Aronow, and Jegga 2009; Navlakha and Kingsford 2010), drug discovery and repurposing (Huang et al. 2013; Scott et al. 2016; Gordon et al. 2020), and vaccine development and optimization (Perrie et al. 2008; Clark et al. 2017). Various *in vitro* and *in vivo* experimental methods detecting the protein-protein interactions have been widely utilized (V. S. Rao et al. 2014). According to the statistics of The Molecular INTeraction Database (MINT), physical association (49.9%) is the top one type of the interactions (Licata et al. 2012). However, we didn't distinguish between different interaction types and interactions detected by different methods in our research, and treated all the experimentally found interactions as the positive samples.

2.1.1 PPI Database Overview

Multiple PPI databases have been created and curated to combine the detected

interactions and served as convenient platforms for researchers to browse and analyze the

interactions and the protein interaction network. MINT, the database of interacting proteins

(DIP), the biomolecular interaction network database (BIND) and Human Protein Reference

Database (HPRD) are PPI databases containing experimentally confirmed interactions, among

which MINT is one of the datasets which has the most curated human PPIs. (Bader, Betel, and

Hogue 2003; Peri et al. 2003; Keshava Prasad et al. 2009; Xenarios et al. 2000; De Las Rivas

and Fontanillo 2010; Licata et al. 2012; Kikugawa et al. 2012; V. S. Rao et al. 2014). In this

research, MINT is one of the sources for our human dataset (see details in Data Preparation).

2.1.2 PPI Prediction Methods Overview

Considering the scale of the possible interaction pairs and the workload of experimental

detection methods, *in silico* PPI prediction has attracted much attention as a prescreening method

based on the known experimentally confirmed PPIs and PPI networks. Such computational

methods can be used to accelerate the process of detecting new PPIs. Different methods have

chosen the information from various aspects to derive the new interactions. Common feature

choices include direct use of sequences, features extracted from sequences, structures, and gene

co-expression correlation (V. S. Rao et al. 2014; Peng et al. 2017).

There are different methods to produce the features, such as calculating from amino acid

sequence with empirical parameters or experimental parameters, multiple sequence alignment,

structure information comparison, gene information collection etc. PrePPI is a structure-based

protein-protein interaction method which predicts a confident score for each protein pair by two

steps: interaction template construction and template scoring (Q. C. Zhang et al. 2012; 2013). At the first step, PrePPI searches for structure-similar proteins for the two interacting proteins and

constructs the interaction templates. Then PrePPI predicts a score for each template from the

structure and property information. Later a newer version of PrePPI was developed with more

evidences: interactions in other databases, expression patterns, partner redundancy and domain

interacting (Garzón et al. 2016). We compared our method to PrePPI on the human dataset in

detail. Though PrePPI offers the interaction template and prediction, the amount of proteins

which have the solved crystal structures or accurate homology models is quite small compared to

the whole protein database, and it is also hard to evaluate how the empirical parameters

contribute to the predictions.

In contrast to protein structures, researchers have easier access to protein sequences and

many feature extraction methods have been developed based on them. Multiple sequence

alignment (MSA) has been widely used to get evolutional information and in protein homology

search. The downsides of MSA are the long running time to calculate the alignment for a large

set of sequences and the weak ability to capture the long-distance dependence.

2.1.3 Sequence Embeddings from Pretrained Language Models

As the development of Natural Language Processing, such techniques have been explored

about their applications in protein sequences. If we treat protein sequences as a language, what

biological information can we learn from it? The Tasks Assessing Protein Embeddings (TAPE)

(R. Rao et al. 2019) and Evolutionary Scale Modeling (ESM) (Rives et al. 2021) have

introduced several semi-supervised learning tasks in protein research domain which embedded

protein sequences by pretrained self-supervised learning networks and then the protein

embeddings were tested on the downstream tasks. Such embeddings generated from the

pretrained models have shown their abilities to increase the performance on almost all the supervised tasks compared to the one-hot encoding inputs and alignment inputs, although the degree of the accuracy improvement varies across different tasks (R. Rao et al. 2019; Rives et al. 2021).

2.1.4 Graph Machine Learning Introduction

There are many naturally existing networking structures in real life, such as webpage network, social network, literature citation network, etc. Such networks are defined as graphs consisting of nodes and edges. The identities such as webpage, people and literature are the nodes, while the links connecting the nodes are edges. The edges can be directed or undirected, and the edges can have different weights as well.

Compared to deep learning methods such as RNNs and CNNs in which the samples are fed into the models with some orders, neural networks on graph can handle the connections between nodes. Machine learning on graph doesn't care about the input order of the nodes, and it updates the hidden representation of a node by calculating the weighted outcome from the neighbors representations (Zhou et al. 2019).

Graph machine learning has attracted attention in chemistry and bioinformatics area. For example, molecules and crystals property prediction (C. Chen et al. 2019), breast cancer type classification (Rhee, Seo, and Kim 2018), etc. Many projects focused on proteins specially, such as protein interface prediction (Fout 2016) and protein-protein interaction prediction with only connection graph (Xiao and Deng 2020).

In our research, we utilized machine learning on graph together with TAPE/ESM protein embeddings to predict PPI interactions.

2.1.5 Data Commons Knowledge Graph Introduction

Data Commons (datacommons.org) is an open knowledge graph repository created and maintained by Google which is trying to combine all the openly available datasets into one large knowledge graph. Data stored in the Data Commons follows the data model in Schema.org where items or entities connect to each other by the connections or properties. The same item and its related information across different databases were cleaned and joined into one entity in the Data Commons, which allows users to easily browse and query the desired objects without efforts to clean data.

In our research, we queried the PPIs in MINT and human protein tissue atlas from Data Commons.

2.1.6 Project Workflow

The workflow of this project consists of two parts: protein embedding generation and classification with supervised deep learning models. We applied the TAPE/ESM pretrained models to generate the protein embeddings which are arrays of length 768/1280 regardless of the protein lengths. On top of the first step of embedding generation we applied the graph neural network to further learn the protein representations from their connections in the protein interaction network. The final input arrays were generated by concatenating the language model embeddings and the learned graph protein representations of the two proteins in each sample. To show how the graph learning step contributes to the prediction, language model embeddings without graph representation learning were directly fed into classification models as well. The last step was to train and test with the classification models. The positive samples have labels as 1 and the negative samples have labels as 0, thus the final supervised classification task is a typical binary classification problem.

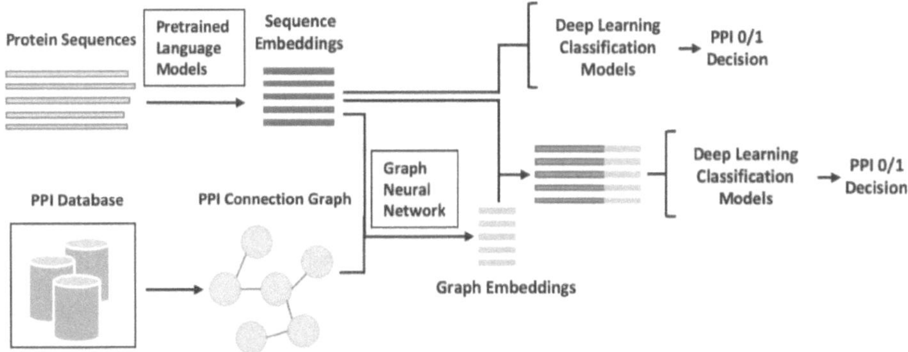

Figure 2.1: Workflow of the PPI prediction in this work. Pretrained language models were used to generate the protein embeddings. Then graph neural network was applied to the embeddings to further learn the protein representations from their connections in the protein interaction networks. The concatenated embeddings from graph embeddings and language model embeddings were fed into classification models to fulfill the PPI prediction task. Language model embeddings alone were used as input to classification models as well.

2.2 Data Preparation

We have trained and tested on two species: *Saccharomyces cerevisiae* (*S. cerevisiae*) and human. *S. cerevisiae* is a species of *Yeast*, and it is a common model organism in experimental biology and bioinformatics (Duina, Miller, and Keeney 2014). As the vaccine development attracts the attention of the world, it is interesting and meaningful to dive in the PPI prediction of human system as well.

PPIs in our research were collected from experimentally confirmed PPI databases, while negative PPIs were paired up with the assumptions that these proteins are not interacting with each other because they are expressed in different tissue cells. To compare with other methods, we chose the datasets that have been tested in the other researchers' work.

2.2.1 *S. cerevisiae* Dataset

The positive PPI dataset we used for *S. cerevisiae* was collected from DIP (Xenarios et al. 2000). This dataset contains paired sequences which have pairwise sequence similarity smaller than 40% (Guo et al. 2008). The final positive PPI dataset has 5,594 interacting protein pairs. The negative PPI dataset was constructed by pairing proteins from different subcellular localizations with the same count as in the positive dataset (Guo et al. 2008; Shen et al. 2007).

There are 2,530 proteins in total which appear as interacting proteins, of which the minimum length is 55 residues while the maximum length is 4,910 residues. The distribution of the *S. cerevisiae* protein lengths is shown in Figure 2.2. We can see that most of the proteins have the lengths less than 2,000 residues. TAPE pretraining BERT model converts each sequence into an embedding matrix. To increase the computation efficiency, we decided to take the average array as the final embedding format, which results in a 768-long array with floating point numbers for each protein. ESM model can take up to 1,022 residues, thus we cut the sequences at 1,022 for it. The average embeddings from ESM have the lengths as 1,280.

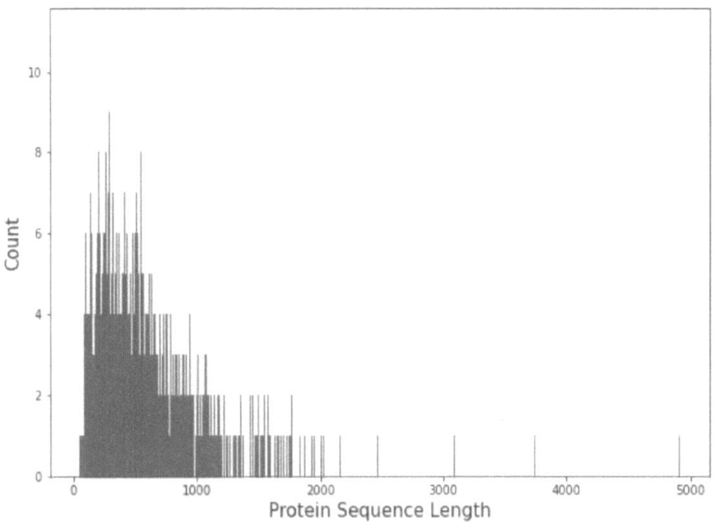

Figure 2.2: Length distribution of proteins in the S. cerevisiae PPI dataset. The maximum length is 4,910 residues and the minimum length is 55 residues. Most of the protein lengths are less than 2,000 residues.

2.2.2 Human Dataset Preparation

To compare the performance of our methods on the human dataset with PrePPI, we held out the positive test dataset (the human HC reference set of which each PPI had at least two literature evidences) which PrePPI used as the positive test set (Garzón et al. 2016). There are 26,983 pairs in the human HC reference set.

For the train positive PPI set, we referred to the MINT (Licata et al. 2012) PPI database and queried the interactions through Data Commons API (datacommons.org). MINT protein-protein interactions have been cleaned and imported to the Data Commons Knowledge Graph. We collected 17,993 protein pairs at this step, of which at least one interacting protein has the labelled species as Homo Sapiens (Human).

13

The negative test set of PrePPI was constructed by randomly pairing proteins and removing the negative PPIs which were experimentally proved to be positive interaction in papers (Garzón et al. 2016). However, we didn't use the same way to construct the negative set. In our research, the reason why we constructed the negative sets for both the train and test datasets differently are as followed. Firstly, the human negative test set of PrePPI contains 1,632,716 proteins pairs, which is a large number compared to the number of protein pairs in the human HC reference set and will need much computing source. The second reason is that we didn't have the list of the human negative test set used by PrePPI neither do we have the corresponding PrePPI scores which were necessary to draw the ROC curve. Thus, we decided to construct a new negative set by pairing proteins expressed in different human tissue-cells.

Proteins expressed in different tissue-cells are considered to have different protein interaction network and many protein interactions are tissue-specific (Bossi and Lehner 2009). Human protein expression profiles were referred to the Human Protein Atlas (Uhlén et al. 2015) and we queried the proteins with its expression level through the open knowledge repository Data Commons API. There were three criteria followed during the negative PPI samples construction.

First criterion was to make sure every negative PPI pair doesn't appear in the positive PPI sets.

Second criterion was to set a standard to define if a protein exists in a tissue-cell or not. For records in the Human Protein Atlas, the protein expression scores include 'not detected', 'low', 'medium' and 'high'. Each record also has its reliability scores derived from the consistency with other evidences, including 'Uncertain', 'Approved', 'Supported', and 'Enhanced', from the weakest level to the strongest level (Uhlén et al. 2015). With the expression scores and reliability levels, we then defined that a protein is not expressed in a tissue-cell when its record has an expression level as 'Not detected' with a reliability level 'Enhanced' and no other records show other expression levels besides 'Not detected'. A protein was defined to be expressed in a tissue-cell when a record shows an expression level as 'high' with a reliability level 'Enhanced'.

The last criterion was to make the contribution of each protein to the negative sets as even as possible (Guo et al. 2008). We set the maximum number of allowed negative samples that one protein can contribute to as 56 and we got 97,226 negative samples at this step.

To compare our results with PrePPI, we would like to have the PrePPI scores for all the PPI samples in the test set. We found that 37,965 out of 97,226 constructed negative samples have the corresponding PrePPI scores in the PrePPI predictions in the file *final_v11.txt* sent by Dr. Donald Petrey from Prof. Barry Honig lab. We retained these 37,965 negative samples in the test set, while kept the remaining 59,261 negative samples in the training set. PrePPI predicts a protein pair as a potential interacting pair if the score is larger than 600 (Garzón et al. 2016). The number of pairs in our test negative set with the PrePPI scores larger than 600 is 384, which is about 1% in the constructed negative set. This low percentage number also supports the rationality of our negative construction strategy.

The train set consists of the MINT positive samples and the constructed negative samples without PrePPI scores. The held-out test set consists of the positive human HC reference set together with the constructed negative samples with PrePPI scores.

2.2.3 Human Dataset Similarity Check

We also researched on the impact of different similarity cutoffs between positive samples and negative samples, and between train set and test set. We calculated the pairwise sequence identity similarity for all the proteins appearing in the human PPI datasets.

Note that we were comparing if two sequence pairs are similar or not instead of directly comparing sequences alone. We assume proteins in one pair are A and B, while proteins in the other pairs are C and D. The comparison process is as follows:

If similarity between A, C and similarity between B, D both are larger than the similarity cutoff, we remove one pair, else we check similarity between A, D and between B, C. If both similarities are larger than the cutoff, we remove one pair. Otherwise the two pairs are considered not similar and both can be kept.

This is to say even if two pairs share the same protein, as long as the two pairs passed the check process above we kept them both.

For the first part of the experiment where we compared the results of language model embeddings alone vs. the concatenated embeddings, we set the train/test similarity cutoff as 25% and BLAST E-value < 10 as default. We also removed the samples in the negative sets by comparing to samples in the positive set with the same similarity criterion, because the positive samples have direct literature evidences of interacting which makes them more reliable when they have conflict labels compared to the similar negative samples.

We got 10,163 train positive samples, 53,306 train negative samples, 29,915 test positive samples and 35,702 test negative samples at the end for the first part experiment.

The second part was to compare the impact of the different similarity cutoffs. The first step in the second part was to remove the similar sequence pairs in negative samples by comparing each negative sample to positive samples in both the train set and the test set. If similar sequence pairs have negative labels and positive labels at the same time, and the confidence of negative labels is lower than that of positive labels which we already know from the construction process of negative samples, the model can be confused when making a prediction. To increase the model prediction ability and decrease the noise from negative sample construction process, we chose 40% and 25% as similarity cutoffs to remove sequences from all the negative samples if they hit at least one sequence pair in the positive sets with the identity similarity higher than the corresponding cutoff.

The second similarity check was to remove the similar sequence pairs between the train set and the test set, to show that the performance of the model comes from the learned internal pattern of different sequences instead of directly mapping to the labels of similar samples. The E-value was set as the default value 10 while the identity similarity cutoffs were set to 15% and 25% respectively. At the end of the data preparation process of the human data similarity check, the sample numbers in each positive, negative, train and test part are shown in Table 2.1.

The length distribution of human proteins appearing in our dataset is shown in the Figure 2.3. The longest length is 34,350 and the shortest length is 24. The proteins with sequence length less than 3,000 residues take up 99.1% of all the proteins, thus we cut the proteins at length 3,000 for those with length larger than 3,000 residues to fit in the computing source for the TAPE model. We also cut the inputs at the length of 1,022 for the ESM model.

neg/pos, train/test	100%, 15%	100%, 25%	40%, 15%	40%, 25%	25%, 15%	25%, 25%
Pos. Train	8410	9098	8566	9326	9775	10558
Neg. Train	33397	35949	33510	36275	34066	36373
Total Train	**41807**	**45047**	**42076**	**45601**	**43841**	**46931**
Pos. Test	29915	29915	29915	29915	29915	29915
Neg. Test	37965	37965	37165	37165	30051	30051
Total Test	**67880**	**67880**	**67080**	**67080**	**59966**	**59966**

Table 2.1: Sample count in each part (positive, negative, train and test) with different sequence identity similarity cutoffs in the human dataset.

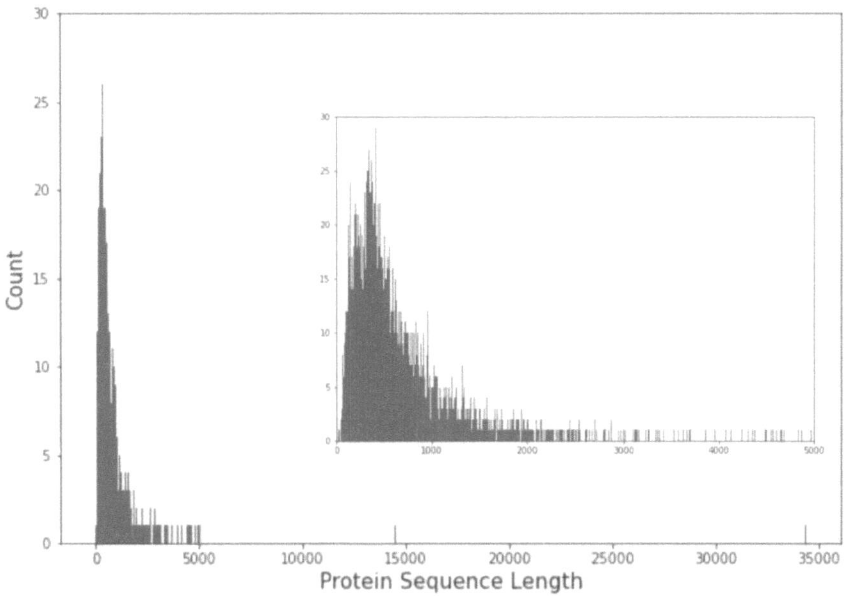

Figure 2.3: Length distribution of proteins in the human PPI dataset. The maximum length is 34,350 residues and the minimum length is 24 residues. 99.1% proteins have length less than 3,000 residues.

2.3 Methods

This method is a semi-supervised learning with the first part as representation learning and the second part as supervised learning.

Many PPI prediction projects were featured in the use of protein structures and homology models, such as MULTIPROSPECTOR (Lu, Lu, and Skolnick 2002), Struct2Net (Singh et al. 2010) and PrePPI (Q. C. Zhang et al. 2013; Garzón et al. 2016). However, the solved protein 3D structures are a small portion of the whole protein database. Currently (in March 2021), RCSB Protein Data Bank (rcsb.org) has 175, 977 solved structures of biological macromolecules according to its homepage. For the protein sequences and functional information database

19

UniPortKB (uniport.org) version 2021_01, its Swiss-Prot database which contains protein records that were literature reported and experts curated, has 564,277 records while the unreviewed database TrEMBL of it has 207,800,733 records, according to the homepage. If we could start with only protein sequences, we would have better chances to find unseen protein interactions. Another advantage of stepping away from the direct use of structure is the less computing cost. It takes many efforts to construct a structure template of high quality if homology modeling is needed during process.

The only data source for our project is the protein sequences, which makes it clean and quick to prepare the datasets. Once trained, we only need the sequences of the two proteins to predict if they are interacting, and it takes less than a second for one prediction.

Did we loss the power of the structure features at all? The answer would be no. The protein sequence search tools such as BLAST (Altschul et al. 1997) and HHblits (Remmert et al. 2012) have been widely used in bioinformatics for calculating sequence similarities and multiple sequence alignment. Many protein properties are conserved during evolution, such as protein motifs, functions and interaction interfaces (Remmert et al. 2012), thus the MSAs should have nice potential to provide useful information for prediction of such properties. An example is that PrePPI applied HHblits on protein sequences to search for the structure templates during the homology modeling process (Garzón et al. 2016). Another example is the success of AlphaFold in protein structure prediction. AlphaFold learns the co-evolutionary information on the sequences and maps the sequences to the contact map (AlQuraishi 2019). These examples show that sequences and their alignment contain critical biological information and have the potential to predict the protein structures. For our method, though we didn't predict protein structures or build homology models as an intermediate step, we actually extracted and applied the evolutional information from sequences by the TAPE/ESM pretraining embedding models.

2.3.1 Protein Sequence Embedding from TAPE/ESM Pretraining Model

TAPE (R. Rao et al. 2019) was pretrained on protein sequences from Pfam (El-Gebali et al. 2019), which contains about 40 million sequence entries. The TAPE self-supervised pretraining tested three different models: LSTM, ResNet, and the BERT model (Devlin et al. 2019) where the BERT model has the best performance. Thus, in our project, we chose the best TAPE pretraining model with BERT as one of our protein embedding method. TAPE has chosen to include the two losses: masked-token prediction (Devlin et al. 2019) and next-token prediction (Peters et al. 2018).

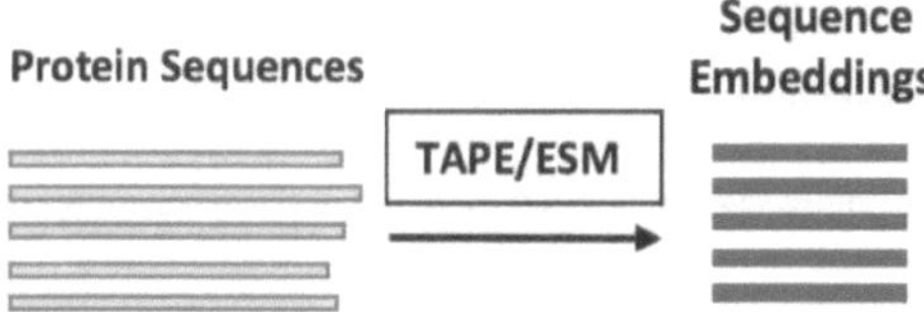

Figure 2.4: Generation of sequence embeddings. Protein sequences of different lengths were embedded by the TAPE/ESM pretrained self-supervised BERT model into array-like embeddings of the same length. The similarity cutoff between train pairs and test pairs in the Human dataset was 25%, while the cutoff was 40% in the *S. cerevisiae* dataset.

ESM was trained on UniParc (The UniProt Consortium 2007), the database that contains the most publicly available protein sequences, which was about 250 million sequences (Rives et al. 2021). It utilized deep Transformers (Vaswani et al. 2017) and the loss used in the masked language modeling (Devlin et al. 2019).

ESM model has 650 million parameters while TAPE model has 38 million parameters, and ESM was reported to exceed TAPE on both the supervised and unsupervised downstream protein related tasks (Rives et al. 2021). We chose the average embeddings over the full embeddings considering the memory limit during the training and less computing time. The average TAPE embeddings have a dimension of 768 while ESM embeddings have a dimension of 1,280. We tested the performance of the both embedding methods in our project.

2.3.2 Graph Embeddings

Different machine learning techniques on graph for various tasks have been developed. Our goal was to learn better representations from the protein interaction network for proteins, and we would have more useful features for the classification model. The model we used in this graph learning part is the attri2vec, which embeds a node's representation together with its neighbors' representation into a new embedding array. Attri2vec uses random walk to collect the neighbor nodes, and the nodes with similar network environment are close in the new embedding space. The learned mapping can be applied to generate the node representations (D. Zhang et al. 2019).

We utilized StellarGraph implementation of attri2vec via its API (CSIRO's Data61 2018).

2.3.3 Classification Models

The language embeddings and language-graph concatenated embeddings were used as inputs to the supervised classification models. We tested the performance of several different models on our dataset including two convolutional neural networks (CNNs), logistic regression and random forest.

CNNs are popular in the image processing tasks such as image classification, image denoising and object recognition , as well as in the natural language processing (Yin et al. 2017). It features the convolutional kernels in the model modules which have the abilities to capture the spatial patterns in the datasets. Compared to the fully connected neural networks, the convolutional layers make the number of parameters fewer. A deeper model requires less parameters to approximate a complex function compared to a wider model, and convolutional modules help the model go deeper (Habibi Aghdam and Jahani Heravi 2017).

Many CNN architectures have been developed with various convolutional blocks and pooling blocks. We tested the models with Inception modules (Szegedy et al. 2014) and ResNet modules (He et al. 2016) in our project.

The Inception-FC model consists of two stacked 1-D Inception modules with dimension reduction adapted from the original 2-D Inception modules (Szegedy et al. 2014) followed by two fully connected layers with units of 128 and 64 respectively. Each dense layer is followed by ReLU activation, a batch normalization layer and a dropout layer. The output layer is activated by sigmoid function to output a value between 0 and 1.

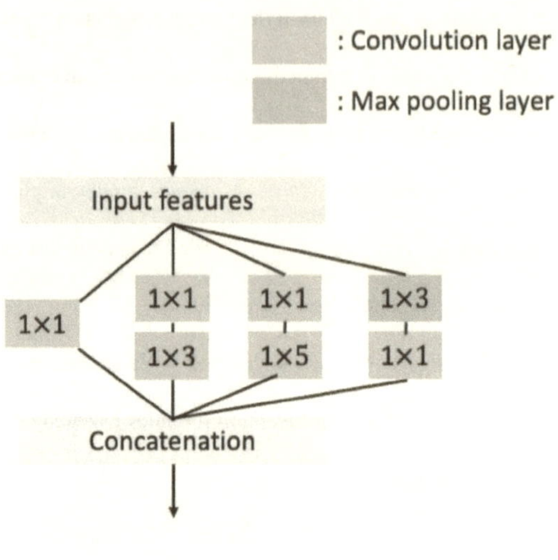

Figure 2.5: Inception module with dimension reductions. The Inception module diagram was adapted from Figure 2 of its original paper (Szegedy et al. 2014).

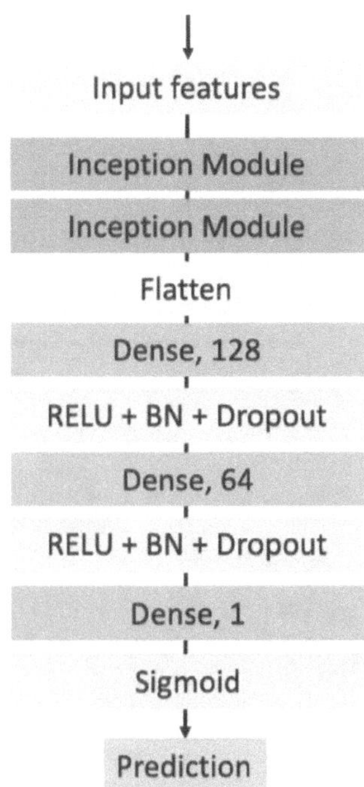

Figure 2.6: Architecture of the Inception-FC model we used in classification. The first two layers are the stacked 1-D Inception modules with dimension reductions, followed by a flatten layer. The flatten layer is followed by two fully connected layers with units of 128 and 64 respectively. Each dense layer is followed by ReLU activation, batch normalization and a dropout layer. The output layer is activated by sigmoid function.

The ResNet-FC model contains five residual network modules followed by one fully connected layer block and an output layer. Each residual module consists of three residual blocks with kernel size 5 and output channel size as 32, 64, 128, 256, 512 respectively. At the end of the last residual network module, a global max layer is followed by the fully connected layers including dense layers of unit 256, 128, and 64 respectively. ReLU is used as the activation

function. The output layer is activated by sigmoid function to output a value between 0 and 1.

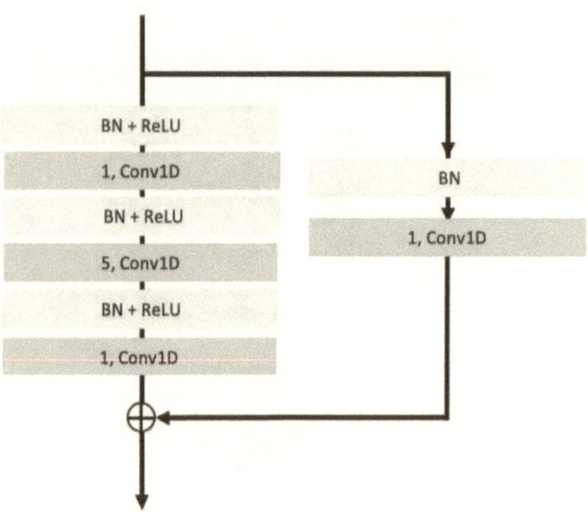

Figure 2.7: Architecture of the residual block. Adapted from Figure 2 in the original paper (He et al. 2016).

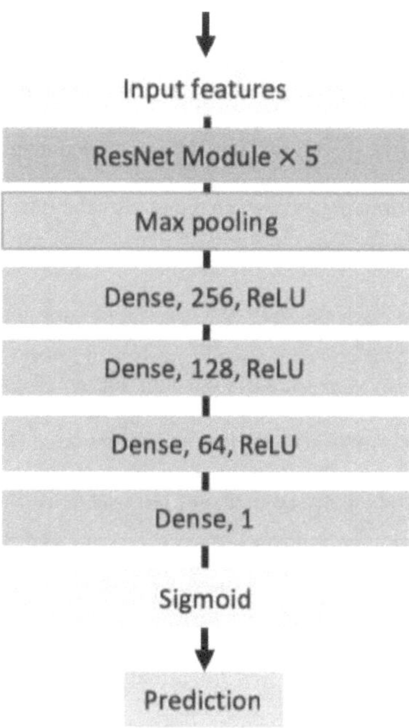

Figure 2.8: Architecture of the ResNet-FC model we used in classification. The input features were fed to the stacked five ResNet modules, of which each contains three residual blocks.

Random forest is an ensemble of decision trees. Samples and features were selected randomly for each decision with bootstrap aggregation approach (Pal 2005; Belgiu and Drăguţ 2016). Compared to decision trees, random forest makes the prediction by the majority vote of the decision tree ensemble and thus reduces the overfitting.

2.3.4 Model Settings for the *S. cerevisiae* Dataset

We ran 5-Fold Cross-Validation on the *S. cerevisiae* dataset for each experiment to test the prediction performance. Specifically, we used the *sklearn* implementation of stratified K-Folds cross-validation data generation method, which guarantees the ratio of positive to negative samples in each fold.

All the positive protein pairs in the train set were treated as the edges of the train graph, and all the protein pairs in the test set were used as the validation set. The training of attri2vec on graph collected samples by random walks and fulfilled an unsupervised learning task. The random walk length, walk number and output dimension were set as 10, 4, 220 respectively. The optimization method was Adam with initial learning rate as 0.0005 while the learning rate reduction (ReduceLROnPlateau) was employed.

For the CNN models, the dropout ratio was 0.6, initial learning rate was 0.001, optimization method was Adam, and the loss was binary cross-entropy.

$$Loss = Y \cdot \left(-\log\left(Y_{pred}\right)\right) + (1 - Y)\left(-\log\left(1 - Y_{pred}\right)\right)$$

Default settings in *sklearn* were applied for logistic regression model and random forest classifier. The average validation results of the five folds were used to compare our results with other methods.

2.3.5 Model Settings for the Human Dataset

We split the dataset into train set and test set, and kept the positive test set the same as the PrePPI's positive test set. We then randomly selected 20% samples out of the train set as the validation set, and used the remaining 80% samples as the final train set. The validation set was used to monitor the training process and select the best model. Then the trained model was evaluated on the test set.

The parameters for graph learning and classification models were the same as used in the *S. cerevisiae* dataset. The evaluation results on the test set were compared to PrePPI results.

2.3.6 Metrics

The evaluation metrics include accuracy (ACC), the area under the ROC curve (AUC), Precision (PRE, also known as positive predictive value), sensitivity (SE, also known as recall or true positive rate (TPR)) and Matthews Correlation Coefficient (MCC) (Du et al. 2017). The Geometric Mean (G-Mean) was used to find the best thresholds (Kubat, Holte, and Matwin 1997). F1-score and MCC are both popular coefficients to evaluate the reliability of the binary classifier, and MCC is thought to be a better one especially in bioinformatics (Matthews 1975; Chicco and Jurman 2020).

Definitions of F1-score, MCC and confusion matrix below refer to (Davis and Goadrich 2006), while other metric definitions refer to (Espíndola and Ebecken 2005).

$$ACC = \frac{TP + TN}{TP + FP + TN + FN}$$

$$PRE = \frac{TP}{TP + FP}$$

$$TPR = Recall = SE = \frac{TP}{TP + FN}$$

$$F1\_score = \frac{2(PRE \times SE)}{PRE + SE}$$

$$MCC = \frac{TP \times TN - FP \times FN}{\sqrt{(TP + FP)(TP + FN)(TN + FP)(TN + FN)}}$$

$$FPR = \frac{FP}{FP + TN}$$

$$Geometric\ Mean = \sqrt{TPR \times (1 - FPR)}$$

Table 2.2: Confusion Matrix

	Predicted Positive	Predicted Negative
Real Positive	True Positive (TP)	False Negative (FN)
Real Negative	False Positive (FP)	True Negative (TN)

2.4 Results

We firstly tested the model with different classification models on the *S. cerevisiae*

datasets with the concatenated graph embeddings and language model pretraining embeddings.

Then we compared our methods to other machine learning methods on the protein-protein

interaction prediction task.

Then we did several experiments on the human dataset and compared the results to

PrePPI (Garzón et al. 2016). We explored the impact of different language modeling

embeddings, graph learning embeddings and different classification models, as well as

experiments on different similarity cutoffs.

2.4.1 Results of the *S. cerevisiae* Dataset

	Precision (%)	Recall (%)	Accuracy (%)
Random Forest	88.63	87.67	88.21
Logistic Regression	82.93	81.75	82.45
Inception-FC	94.27	90.68	92.58
ResNet-FC	95.14	92.31	93.78

Table 2.3: Prediction results from different classification models on the *S. cerevisiae* dataset. The inputs were the concatenation of graph embeddings and the TAPE pretraining embeddings. Results are from the average value out of the 5-fold cross validation.

The input features for the classification models were the concatenated embeddings from

the graph embeddings and TAPE pretraining embeddings. Results have shown that the deep

learning models have better results than Random Forest and Logistic Regression models. Within

the two CNN models, the result from the ResNet model is better than that from the Inception

model.

	Precision (%)	Recall (%)	Accuracy (%)
Guo's work (Guo et al. 2008)	88.87	89.93	89.33
You's work (You et al. 2013)	87.59	86.15	87.00
You's work (You et al. 2014)	91.94	90.67	91.36
DPPI (Hashemifar et al. 2018)	**96.68**	**92.24**	**94.55**
DeepPPI (Du et al. 2017)	**96.65**	**92.06**	**94.43**
TAPE_inception	93.68	90.00	92.04
graph_TAPE_concat_inception	94.27	90.68	92.58
TAPE_resnet	95.15	92.31	93.80
graph_TAPE_concat_resnet	95.14	92.31	93.78
esm_inception	94.25	90.98	92.71
graph_esm_inception	94.81	92.06	93.51
esm_resnet	95.49	92.70	94.16
graph_esm_resnet	**95.62**	**92.68**	**94.21**

Table 2.4: Our results with different experimental settings and the comparison to other sequence-based methods on the *S. cerevisiae* dataset. The first four methods are sequence-based while DeepPPI has utilized other features besides protein sequences.

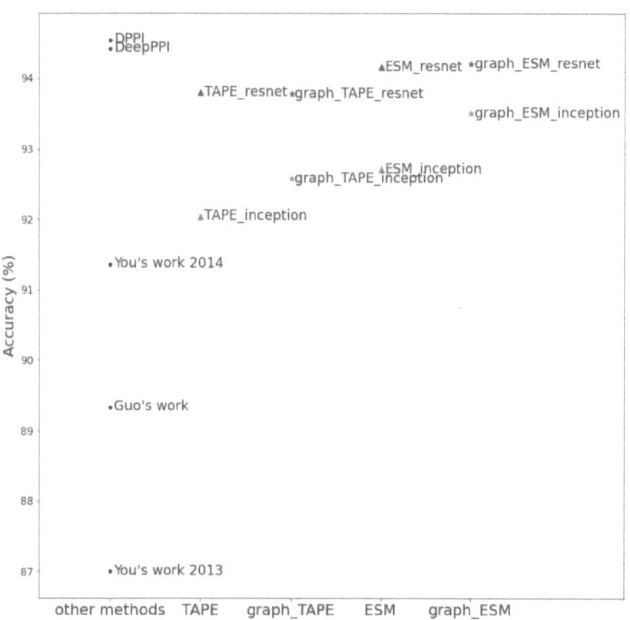

Figure 2.9: Accuracies at thresholds of our experiments with different settings and the accuracies of other methods on the *S. cerevisiae* dataset.

We compared our methods to other sequence based methods including Guo's work (Guo et al. 2008), You's work in 2013 (You et al. 2013), You's work in 2014 (You et al. 2014) and Hashemifar's work (Hashemifar et al. 2018). We also compared our methods to Du's work (Du et al. 2017) which utilized other complex empirical parameters. Currently our best method is only slightly worse than DPPI (Hashemifar et al. 2018) and exceeds other sequence-based methods.

2.4.2 Results of the Human Dataset

For the comparison on the human dataset between our methods and PrePPI, we drew the

Receiver Operating Characteristic (ROC) curves for the results on test set. ROC curve was

plotted with TPR vs. false positive rate (FPR, or fall-out). Data of PrePPI was referred to the

2016 PrePPI paper (Garzón et al. 2016) and the PrePPI scores received from Dr. Donald Petrey.

AUC ranges from 0 to 1 and the classifier is believed to have better classification performance

when the AUC value is closer to 1.

In the first part we compared the TAPE/ESM embeddings alone to the concatenation of

TAPE/ESM embeddings and the graph embeddings, using the two CNN models. In the second

part we explored the impact of different similarity cutoffs.

The results of the first part are shown in Figure 2.10 and Table 2.5. According to the

ROC curves, our classifiers have better TPRs starting from very low FPR. Our classifiers have

better AUCs and accuracies at the best thresholds compared to PrePPI as well.

Most results from the concatenation of graph embeddings and language model

embeddings are slightly worse than the corresponding results of the language model embeddings

alone, which is contrary to the trend in the *S. cerevisiae* dataset. The ratio of graph edges to

graph nodes is calculated as 1.054 for the human dataset, and 2.210 for the *S. cerevisiae* dataset.

Thus, the connection sparsity in the human PPI connection map is considered as a reason why

graph learning doesn't show a good impact in most cases in the human dataset. The best result of

our methods comes from the concatenation of ESM embedding and the graph embeddings with

the ResNet classification model, which has an accuracy of 89.80% at the best threshold and an

AUC of 0.95, while the accuracy at the best threshold and the AUC of PrePPI are 85.80% and

0.91.

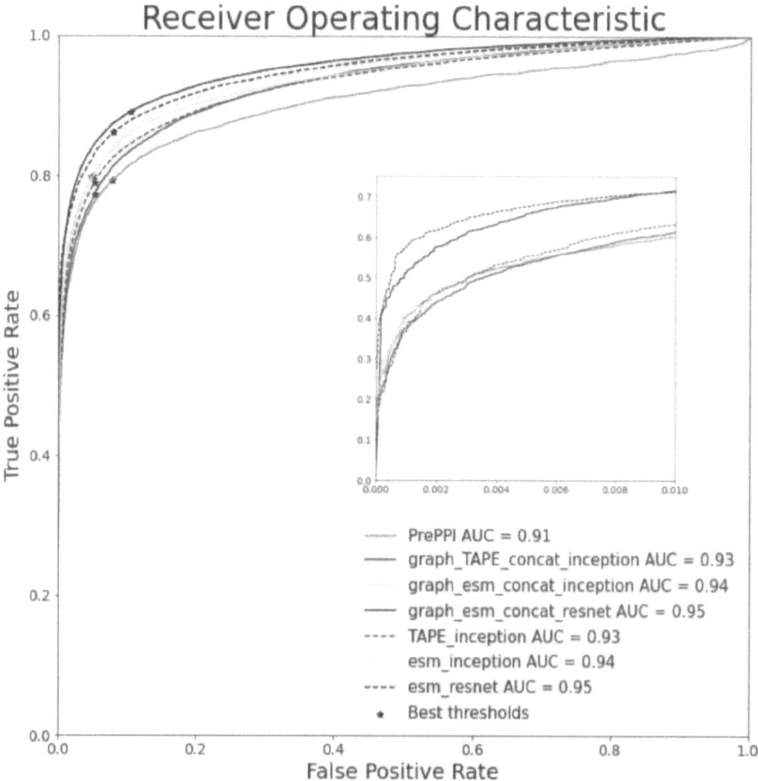

Figure 2.10: Receiver Operating Characteristic (ROC) Curve for the human test set, with different embedding methods. The dataset has a similarity cutoff of 25% between the negative and the positive samples, and a 25% similarity cutoff between the train and the test samples. PrePPI data is from the 2016 PrePPI paper (Garzón et al. 2016) and PrePPI scores received from Honig lab.

	Best threshold	G-Mean	ACC (%)	PRE (%)	SE (%)	F1-score	MCC
PrePPI	26.68	0.855	85.80	86.41	85.78	0.8574	0.7219

TAPE Inception	0.001689	0.872	87.34	87.62	87.33	0.8731	0.7494
graph_TAPE Inception	0.001144	0.866	86.74	87.06	86.73	0.8671	0.7379
ESM Inception	0.00212	0.882	88.20	88.27	88.19	0.8819	0.7646
graph_ESM Inception	0.00279	0.879	87.92	88.04	87.92	0.8791	0.7596
TAPE ResNet	7.17E-06	0.892	89.25	89.34	89.25	0.8925	0.7859
graph_TAPE ResNet	1.32E-06	0.891	89.11	89.23	89.10	0.8910	0.7833
ESM ResNet	1.71E-04	0.891	89.16	89.40	89.15	0.8914	0.7854
graph_ESM ResNet	7.35E-07	0.898	89.80	89.90	89.79	0.8979	0.7969

Table 2.5: Results of human test sets at the best thresholds from the experiments using language embeddings alone or the concatenation of language embeddings and graph embeddings. ResNet and Inception networks were used as classification models. PrePPI data is calculated from the data in 2016 PrePPI paper (Garzón et al. 2016) and PrePPI scores received from Honig lab.

The results of the second experiment are shown in Figure 2.12 and Table 2.6 which explored the impact of different similarity cutoffs. The best threshold for each method was found by maximizing G-Mean since our dataset had imbalanced number of positive samples and negative samples. The results of each experiment with the corresponding thresholds are shown in Table 2.6.

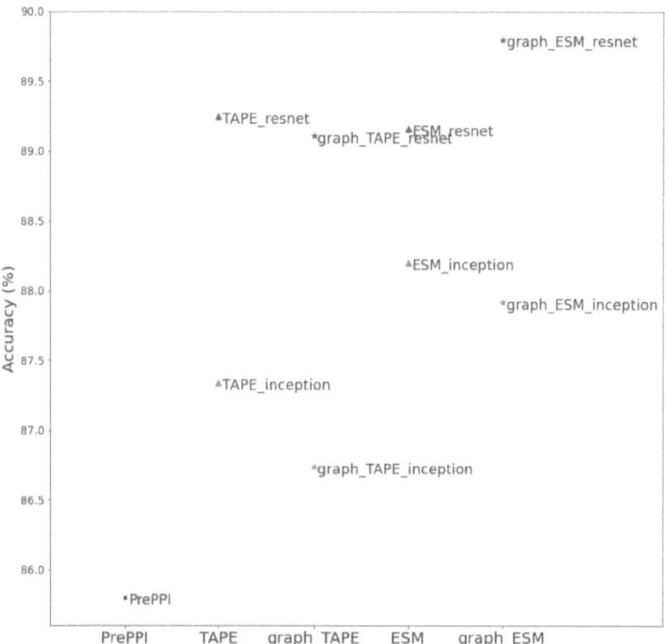

Figure 2.11: Accuracies of different models on human dataset compared to PrePPI accuracy. PrePPI data is calculated from the data in 2016 PrePPI paper (Garzón et al. 2016) and PrePPI scores received from Honig lab.

Experiments shown as dashed lines share the same test set where the similarity cutoff is

40% which was used to remove the similar sequence pairs from the negative set by comparing to

the positive set, while the experiments shown as solid lines share the same test set with the

similarity cutoff as 15%. Experiments shown as dotted lines didn't remove any similar sequence

pairs between negative pairs and positive pairs, of which the cutoff can also been seen as 100%.

Because each negative/positive cutoff produces one test set, we analyze them as three groups.

The dashed purple line has a negative/positive similarity cutoff as 40% and a train/test similarity cutoff as 15%, which means no similar sequence pairs with similarity larger than 40% were labelled as 1 and 0 at the same time, and no sequence pairs with pair similarity larger than 15% exist in the train set and test set at the same time. The dashed blue line has the same negative/positive similarity cutoff as 40% while the train/test similarity cutoff is 25%. The dashed red line was drawn from the PrePPI scores with the same test set. The AUCs of the two lines from our results are both 0.92, which are higher compared to the PrePPI AUC of 0.89. This shows that our classifiers have better overall performance than PrePPI. The accuracies of predictions at the corresponding best thresholds are 85.03%, 85.42% and 86.04% for PrePPI, our method with the train/test similarity cutoff as 15% and our method with the train/test similarity cutoff as 25%, respectively, as shown in Table 2.6. Our methods also have better MCCs and F1 scores compared to PrePPI, showing the higher quality of our classifiers.

PrePPI has a better TPR than our methods when FPR is less than about 0.01. When FPR is larger than 0.01, our method with the train/test similarity cutoff as 25% has a higher TPR than PrePPI, while when FPR is larger than 0.04, our method's TPR with the train/test similarity cutoff as 15% exceeds PrePPI's TPR. This is to say our methods have better abilities to recognize true positive samples when we could have a little tolerance of the FPR.

In the group with negative/positive similarity cutoff as 25% shown with solid lines, the results compared to PrePPI are similar to the experiment with negative/positive similarity cutoff as 40%. At the best thresholds, the accuracies are 85.80%, 86.14% and 87.34% for PrePPI, our method with 15% train/test similarity cutoff and our method with 25% train/test similarity cutoff.

When the FPR is larger than 0.003, the TPR of 25% similarity cutoff experiment starts to exceed the PrePPI TPR, while when the FPR is larger than 0.008 the TPR of 15% similarity cutoff experiment is better than the PrePPI TPR.

In the group where we kept the similar samples in the positive and negative samples, the results are similar to what we have discussed above as well. The accuracies are 84.80%, 84.93% and 85.31% respectively for PrePPI, our method with 15% train/test similarity cutoff and our method with 25% train/test similarity cutoff. The TPR of our method with 25% similarity cutoff starts to exceed the TPR of PrePPI at 0.005 FPR and the TPR of our method with 15% similarity cutoff becomes better than the TPR of PrePPI at about 0.016 FPR.

Comparing between our two classifiers with train/test similarity cutoff as 15% and 25% in each group, though they have similar AUCs, we can see that the classifier with 25% similarity cutoff has a better performance compared to the classifier with 15% similarity cutoff. This is understandable that when the similarity between the test set and train set is higher, the performance of the classifier on the test set is higher. However, 25% identity similarity is already a reasonable cutoff, while 15% is a very strict setting. The good performance of our classifiers even with 15% cutoff shows that our model has the ability to capture the hidden pattern underlying the sequences and the trained classifier has reliable prediction results on protein-protein interaction task.

Comparing the classifiers across the three groups by the ROC curves and the metrics, we can rank the performance from the lowest to the higest as this: group with negative/positive similarity cutoff as 100%, 40% and 25%, respectively. The accuracies of our methods at the best thresholds are 84.93%, 85.31%, 85.42%, 86.04%, 86.14%, 87.34% from the top to the bottom rows in Table 2.6. The corresponding train positive counts are 8410, 9098, 8566, 9326, 9775 and

10558. Though the size of negative set should have some impact, we think that the quality and size of the train positive samples play an important role on the model performance since the test accuracies and train positive counts share the same trend.

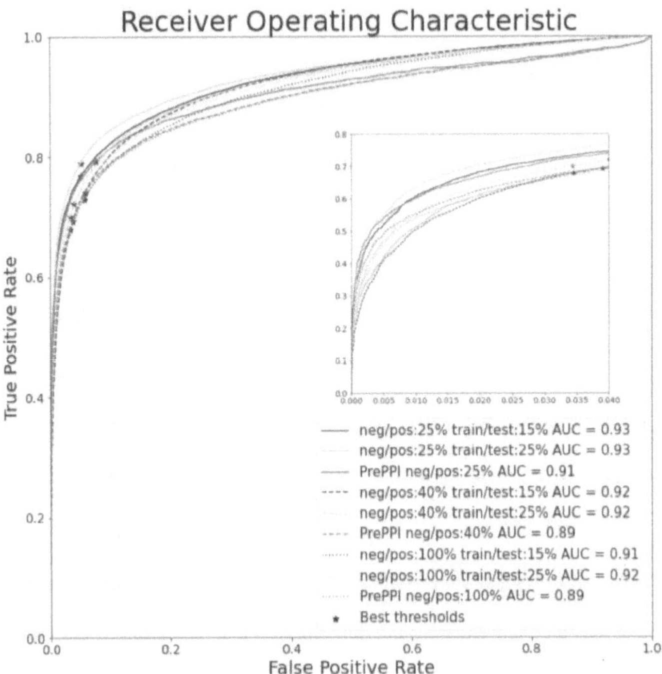

Figure 2.12: Receiver Operating Characteristic (ROC) Curve for the human test set with different similarity cutoffs. PrePPI data is from the 2016 PrePPI paper (Garzón et al. 2016) and PrePPI scores received from Honig lab.

40

	Best threshold	G-Mean	ACC (%)	PRE (%)	SE (%)	F1-score	MCC
PrePPI neg/pos:100%	28.18	0.840	84.80	84.91	84.18	0.8444	0.6908
neg/pos:100% train/test:15%	0.002298	0.840	84.93	85.04	84.31	0.8457	0.6934
neg/pos:100% train/test:25%	0.001235	0.848	85.31	85.26	84.86	0.8502	0.7012
PrePPI neg/pos:40%	28.18	0.840	85.03	85.24	84.44	0.8471	0.6967
neg/pos:40% train/test:15%	0.001816	0.850	85.42	85.37	85.06	0.8519	0.7043
neg/pos:40% train/test:25%	0.001437	0.855	86.04	86.07	85.62	0.8579	0.7169
PrePPI neg/pos:25%	26.68	0.855	85.80	86.41	85.78	0.8574	0.7219
neg/pos:25% train/test:15%	0.001323	0.860	86.14	86.55	86.13	0.8610	0.7268
neg/pos:25% train/test:25%	0.001689	0.872	87.34	87.62	87.33	0.8731	0.7494

Table 2.6: Results of human test sets at the best thresholds from the experiments with different cutoffs between the train and test set, and the negative and positive set. PrePPI data is calculated from the data in 2016 PrePPI paper (Garzón et al. 2016) and PrePPI scores received from Honig lab.

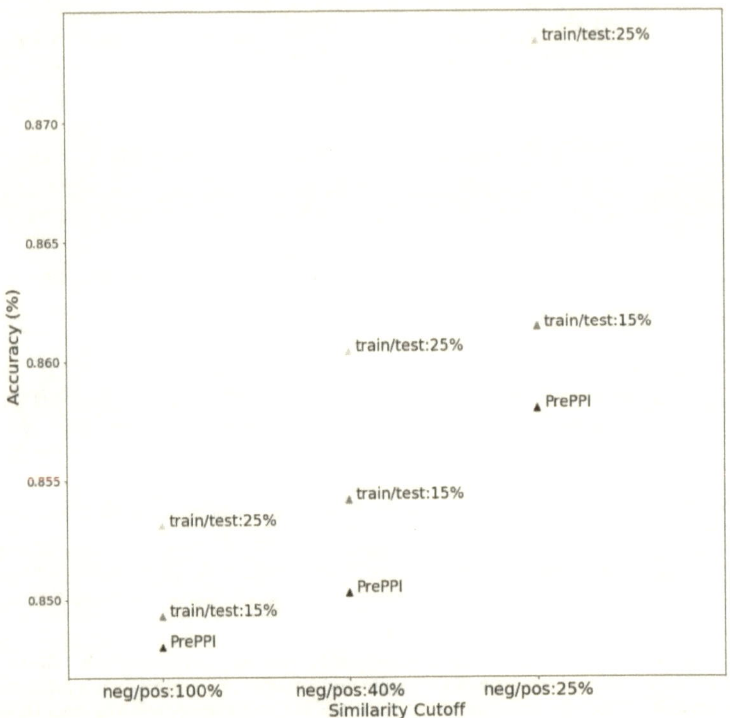

Figure 2.13: Accuracies of human test sets at the best thresholds with different similarity conditions. PrePPI data is calculated from the data in 2016 PrePPI paper (Garzón et al. 2016) and PrePPI scores received from Honig lab.

2.4.3 Discussion

The inputs in our method are the directly learned embeddings from protein sequences, which are different from the inputs from multiple sequence alignment and empirical parameters in some other sequence-based methods. The pretrained embedding models were built with

Natural Language Processing techniques and our results have shown the success of transferring such techniques from learning human language to protein sequences. The training process of such embedding models is time consuming and needs quite a bit GPU resource. However, once trained, the embedding producing process is quite fast, especially compared to the multiple sequence alignment process. In our research, we didn't train such embedding models ourselves but applied the pretrained language models. TAPE/ESM embeddings have been proven that they can increase the accuracies of some downstream biological tasks compared to using one-hot sequence encoding and multiple sequence alignment. Our research has shown that such embeddings learned directly from protein sequences can be used to predict protein-protein interactions, and we have compared our results to other methods which utilized features beyond sequences and alignment.

We tested our methods on the *S. cerevisiae* dataset with several classification models including Random Forest, Logistical Regression and deep learning models (CNN layers with fully connected layers). The best accuracy of our methods comes from the concatenated embeddings of ESM embeddings and graph learning embeddings with a CNN model (ResNet layers with fully connected layers), which is 94.21% and better than four other methods that we compared to. The best accuracy of other methods is 94.55% from DPPI (Hashemifar et al. 2018). Though currently our method is not better than it, we still have several advantages. DPPI utilizes multiple sequence alignment for each new sample, which is very time consuming. Secondly, there are still many deep learning model architectures that we haven't tried yet, which have the chance to result in higher accuracies. Lastly, different language models are currently being applied to different protein databases, which could produce better embeddings as well.

Then we tested our methods on the human dataset and compared to PrePPI. For all the comparison, we kept the test set of our methods and PrePPI the same in each group. We firstly tested the impact of adding graph learning embeddings, and the results show that the graph embeddings learned from the interaction map couldn't help the prediction much. The average ratio of edges to nodes is only 1.054 for the human dataset, which means each protein participates in about two interactions. However, each protein in the *S. cerevisiae* dataset participates in about 4.4 interactions. Thus, we suspect that the human protein interaction connection map is too sparse for the model to learn the useful features. For the second experiment with different similarity cutoffs, we only used the TAPE embeddings which saves the time cost for the graph learning process.

For the similarity cutoff experiments, our results show that our classifiers have better performance than PrePPI. Compared to the large feature collection that PrePPI used including structure information, our methods are faster at the input preparation step. We can also apply our method to a larger possible protein interaction space, since PrePPI is limited to the proteins with solved structures but our methods only need the protein sequences, though with the shortcoming that we didn't provide the interaction structure model as PrePPI did.

In this research, we only trained on the human interactions from MINT while excluding the interactions in the human HC reference set which were used as the test set in PrePPI. MINT PPI collection is a small part of the available human PPIs and a better prediction model should be achieved if we could include more positive samples from different PPI datasets.

PrePPI was trained on the yeast dataset and tested on the human dataset, while our method was trained and tested on the human dataset. Also note that when the embedding model was trained, proteins of all species were combined and not distinguished. It is reasonable to assume that we could also train on other species to increase the size of the train set, and fine-tune our model on the human dataset to achieve a better accuracy.

Chapter 3 Predicting Inter-residue Distances in Protein Loops with Deep Learning and Its Incorporation to PLOP for Loop Reconstruction

This chapter is the draft paper of the research which applies deep learning techniques to the distance map prediction in the protein loops and then reconstructs the full-atom loop structure. We used two different strategies to reconstruct the loops, one is to reconstruct loops solely, with Multidimensional Scaling (MDS) and PULCHRA; the other one is to reconstruct loops within the whole protein structure by combining PLOP (Jacobson et al. 2004) and the predicted distance maps.

3.1 Introduction

To make the protein loop structures *in silico* more precise, many efforts have been made to homology loop modeling and *de novo* loop modeling, however there is always a trade-off between the accuracy and the cost. A fast and accurate loop modeling method can give the downstream protein simulation tasks more accurate starting structures.

Template-based methods such as LoopIng (Messih, Lepore, and Tramontano 2015) and DeReUS-Loop (Karami et al. 2018) are limiting the conformations of their predicted loops to the known protein structure database. Considering that loop conformations in the homologs are less conservative, *ab initio* or hybrid methods can have more flexible loop modeling results, such as HLP-SS (Sellers et al. 2008; Jacobson et al. 2004), Rosetta-NGK (Stein and Kortemme 2013), GalaxyLoop-PS2 (Li et al. 2017), RCD+ (López-Blanco et al. 2016) and Sphinx (Marks et al. 2017). However, the common problem of *ab initio* methods is the large computing cost which may take hundreds of hours, especially when it comes to the loops longer than 12 residues.

Considering the efficiency, a trained machine learning model can generate a new prediction fast because the parameters have been settled and thus is a good choice to be used as a low-cost computing method. We utilized a deep learning model to predict the distance map of C_α in the loop area. We then used two methods to reconstruct the full-atom loop structures, one is PULCHRA (Rotkiewicz and Skolnick 2008) while the other one is the PLOP program (Jacobson et al. 2004). The latter one reconstructs the loop structure within a whole protein, which is more useful in the protein modeling than only predicting the loop fragments. The combination of machine learning and this *ab initio* loop modeling method achieves a higher accuracy with a lower computing cost.

3.2 Datasets

We used the loops extracted from the whole PDB database, with 90 percent identities. We assumed that the loops are less conserved within the homologous proteins than other secondary structures, thus we set the identity at a relatively high value. To align the predicted loops at the gap region, one more residue as anchor at each end of the loop needs to be predicted as well. Besides the loop sequence, we consider the 8-residue extra length at each end as the source for environment information. For example, the input sequence length is 24 residues for 8-residue loops while the input length is 28 residues for 12-residue loops.

We firstly checked the integrity of the dataset. The primary sequence information of the loop plus the two flanks should be complete as well as the loop coordinates in a PDB file. Also, the loops of which the flanks include another loop were excluded. The loops with the same sequence were removed as well. To compare with other methods, the related bench loops were retained separately.

For the loop fragment reconstruction experiment by PULCHRA, we used the CASP11 and CASP12 loop sets from DaReUS-Loop (Karami et al. 2018) as the test set and calculated the local RMSD results.

We then reconstructed the loops combining PLOP and the machine learning predicted distances by using the standard 8 and 12 residues benchmarks. (Park et al. 2014; López-Blanco et al. 2016)

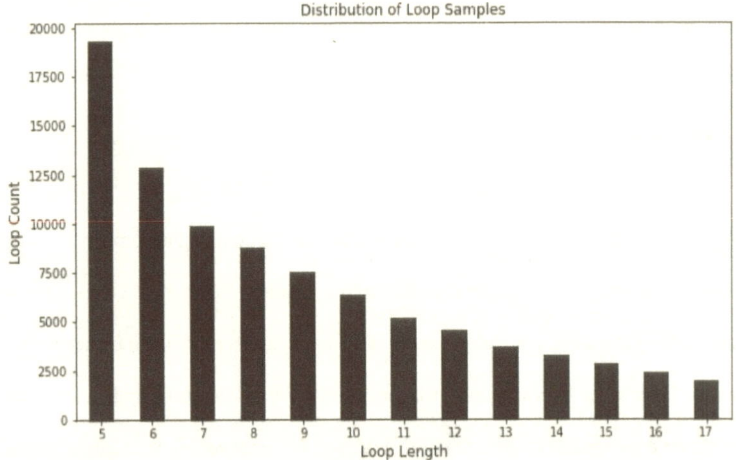

Figure 3.1: Distribution of loop lengths in our loop samples.

3.3 Methods

Error! Reference source not found. shows the flowchart of our work. An input consists of t he primary sequence of the loop region and the 8-res flanks at each end. Then the amino acid sequence is encoded into a list of unique indices and fed to the model which outputs the predicted inter-residue distances of C_α coordinates in the loop region. The final loop reconstruction is done by PULCHRA (loop fragment alone) and by PLOP with the predicted

distances as constraints (whole protein model with loop reconstructed).

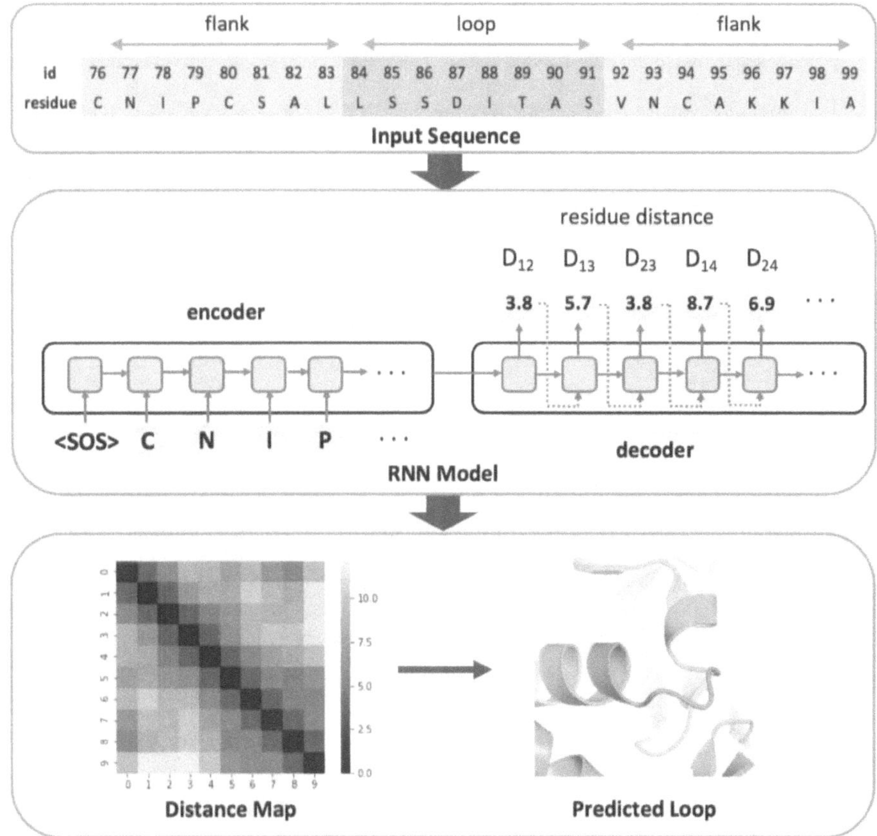

Figure 3.2: Flowchart of the work. The primary sequence inputs were fed into the RNN model to predict the inter-residue distances of the loop Cα coordinates. Then the loops were reconstructed by PULCHRA or PLOP program.

3.3.1 RNN Model and Attention Mechanism

We used the seq2seq model (Sutskever, Vinyals, and Le 2014) with Luong global attention mechanism (Luong, Pham, and Manning 2015) to learn the predictions from loop primary sequences to the inter-residue distances. We used the Gated Recurrent Unit (GRU) (Cho et al. 2014) for both the encoder and decoder networks. GRU is a variant of Long short-term memory (LSTM) (Hochreiter and Schmidhuber 1997), with a simpler structure but comparable results (Chung et al. 2014).

The final model architecture is the multi-layer bidirectional RNN with attention. We tested the prediction abilities of models with different layer scales.

3.3.2 Training Loss

Cross-entropy loss was used in the training process where the distance values in the output were treated as discrete variables with difference of 0.1. To accelerate the convergence, we used Teacher Forcing technique that uses the ground truth replacing the actual output of each unit during the training process (Williams and Zipser 1989). Samples of different lengths were trained separately to have less training cost and the desired predicting length of the distance array.

During the training, the loss takes the form of multi-class cross entropy, and the loss for one distance is:

$$- \sum_{i=1}^{c} y_{i,d} \log (p_{i,d})$$

If it predicts the distance d with the correct index, the $y_{i,d}$ is 1, otherwise $y_{i,d}$ is 0; $p_{i,d}$ is the probability of predicting distance d as index i. The total loss is the average of all the distance losses. Root Mean Square Deviation (RMSD) was used to show the deviation of the predicted distances from the distances calculated from the PDB files.

3.3.3 Models

We implemented our models according to the seq2seq network (Sutskever, Vinyals, and Le 2014), Luong attention mechanism (Luong, Pham, and Manning 2015) and the practical machine translation example: Translation with a Sequence to Sequence Network and Attention (pytorch.org/tutorials).

We split the dataset into train, validation and test with the ratio of about 4:1:1, and utilized early-stopping for the training process.

The loop length range of our samples is from 5 residues to 17 residues. Considering the training cost, we built separate models for loops of each length, even if seq2seq models have the ability to handle inputs of different length (Sutskever, Vinyals, and Le 2014). The number of encoder and decoder layers has the most impact on the accuracy, and the best layer numbers for the different models together with the RMSD of the distance predictions for ML_PULCHRA and ML_PLOP are shown in Table 3.1 and Table 3.6.

3.3.4 Impact of Teacher Forcing Ratio

To decide the best teacher forcing ratio to use in our model, we tested the ratios of 0.5, 0.7, 0.8 and 1.0 on the models for loops with the length of 12 residues. Other parameters included layer number as 4, learning rate as 0.001, batch size as 32, dropout ratio as 0.3 and hidden size as 512. The best result came from the experiment with teacher forcing ratio as 1 as shown in Figure 3.3. It means when we fed the ground truth label from the previous unit into the next unit for every unit, our model converged at the fastest pace. The teacher forcing ratio was set to 1 for all the models of different lengths for later experiments.

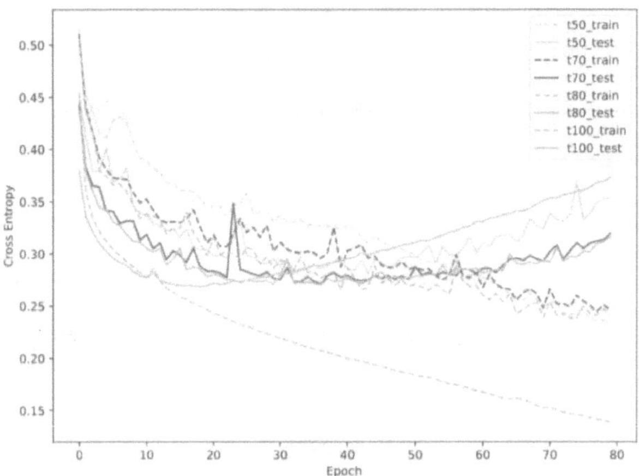

Figure 3.3: Cross entropy loss in training with different teacher forcing ratio of 12-res model.

3.3.5 Multidimensional Scaling and PULCHRA

Sklearn.manifold.smacof function which implements metric multidimensional scaling was used to reconstruct the coordinates from the distance matrix (Borg and Groenen 2005).

PULCHRA, which was developed in 2008, is a fast reconstruction method for full-atom protein models. PULCHRA can start from the C_α positions and then add the backbone and side chain atoms at the appropriate positions to minimize the energy while forming the hydrogen bonds and decreasing the clashes during the building process (Rotkiewicz and Skolnick 2008). This is suitable for our task which is to quickly reconstruct loops from the C_α coordinates.

3.4 ML_PULCHRA Results

3.4.1 Inter-Residue Distance Prediction Accuracy

Loop Length	Best Model Layer	Test Size	Distance Test RMSD
4	3	2496	0.60
5	3	1960	0.63
6	5	1302	0.70
7	5	1008	0.75
8	5	902	0.74
9	7	783	0.73
10	6	648	0.76
11	8	521	0.78
12	8	459	0.79
13	9	371	0.77
14	8	328	0.79
15	9	281	0.83
16	9	235	0.84
17	10	194	0.87

Table 3.1: Number of layers, test sample size and test sample RMSD for the distance prediction of the best model of each loop length.

The layer number for the best model at each loop length and the corresponding RMSD

are shown in Table 3.1. The loops in the test set were removed from the training set for the

ML_PULCHRA experiment.

To evaluate the performance of our models at predicting the inter-residue loop distances, we calculated the Pearson correlation coefficients in Figure 3.4. Pearson correlation examines the linear relation between two variables, while 1 means the perfect positive linear relation, -1 means the perfect negative linear relation and 0 means no linear relation (Benesty et al. 2009). We combined the data from all the models for different loop lengths from 4 to 17 residues. When the two residues are adjacent, the interval length is 1. The coefficients of the interval length longer than 2 are larger than 0.9 showing that there is great correlation between the predicted distances and the calculated true distances from the PDB files.

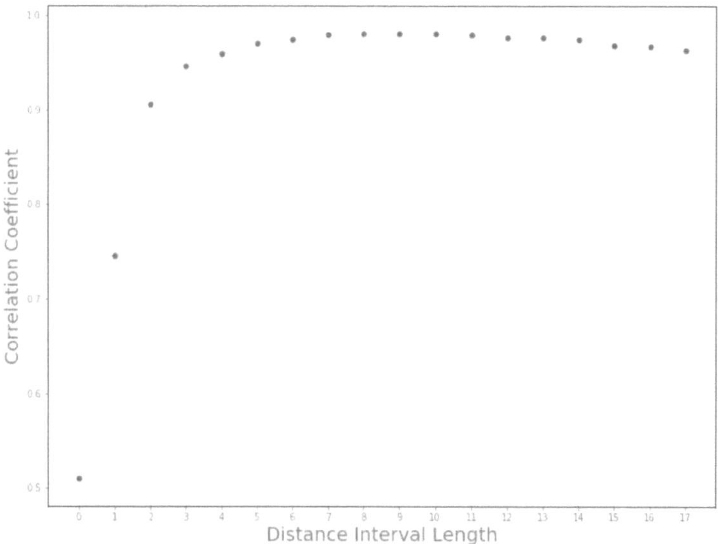

Figure 3.4: The Pearson correlation coefficients of the inter-residue distance prediction at each distance interval length of the combined results from all the models.

The Pearson coefficient for the distances with interval length as 2 is around 0.75 and for the distances with interval length as 1 is 0.51. We think that this is because the linear relation is not that obvious between the RMSD and the small distances when the noise fluctuations can have a large impact. This can be further confirmed that some outliers will have a large impact, though the absolute RMSDs are very small for the predicted distances with interval length smaller than 3, as shown in Figure 3.5.

The RMSDs of the predicted distances compared to the ground truth are less than 2Å except for the outliers at the loop length of 15 residues from Figure 3.5. For the distances with the interval length less than 5 residues, we can see that all the models have similar RMSDs while the model with a larger loop length has a lower RMSD when the interval length gets larger. The reason is that the models for the longer loop length are more complex, and thus have better abilities to predict the distances of residue pairs with long interval length.

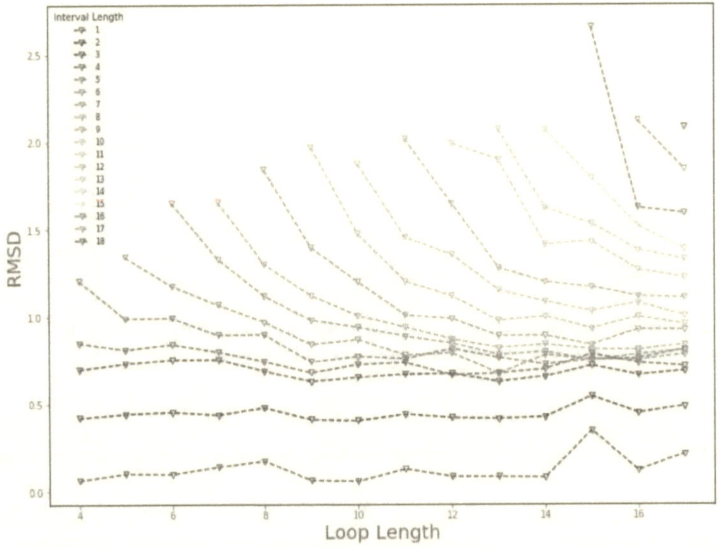

Figure 3.5: RMSD of the inter-residue distances for each residue interval length from each model at different loop length

3.4.2 Loop Reconstruction Results Compared to Other Methods

To directly compare the results of our method to other loop modeling methods, MDS and

PULCHRA were used to reconstruct loops from the predicted distances. For CASP11 and

CASP12 loops, we compared the local RMSD of our results with *ab initio* methods NGK (Stein

and Kortemme 2013), GalaxyLoopPS2 (Park et al. 2014), template methods LoopIng (Messih,

Lepore, and Tramontano 2015), DaReUS-Loop (Karami et al. 2018), and a combined method

Sphinx (Marks et al. 2017). The CASP11 and CASP12 datasets, and the local RMSD results of

these methods were taken from the supplementary files of the DaReUS-Loop (Karami et al.

2018). The results of our method SeqDist come from the average values out of 20 reconstruction

runs.

Our RMSD result is 1.79 for the CASP11 dataset, which is the second lowest average

local RMSD, only higher than that of the Galaxy-PS2's result, 1.67. Also, our method SeqDist

has the lowest standard deviation as 0.41. For CASP12, our method has the lowest average local

RMSD as 1.75, with an outstanding standard deviation as 0.41. These results have shown that

our method has the outstanding abilities to predict the loop local conformation, without

considering the whole protein structure.

	Galaxy PS2	NGK	Sphinx	LoopIng	DaReUS-Loop	**SeqDist**
AVG.	1.67	1.94	4.31	2.61	2.36	1.79
STD.DEV	1.07	1.21	3.31	1.19	1.29	0.41

Table 3.2: CASP 11 Results for the local loop prediction. The data of other methods is taken from DaReUS-Loop (Karami et al. 2018).

	Galaxy PS2	NGK	Sphinx	LoopIng	DaReUS-Loop	**SeqDist**
AVG.	1.96	1.98	4.40	2.64	2.22	1.75
STD.DEV	1.26	1.21	3.06	1.42	1.03	0.41

Table 3.3: CASP 12 Results for the local loop prediction. The data of other methods is taken from DaReUS-Loop (Karami et al. 2018).

To further compare with a deep learning method called Exp_GAN (Li et al. 2017, 201), we applied our method to two bench sets, one contains 19 loops with the length of 8 residues and the other contains 19 loops with the length of 12 residues, of which the protein backbones were perturbed as described by Park *et al.* (Park et al. 2014). Results of NGK, GalaxyLoopPS2 and Exp_GAN were taken from Li's work (Li et al. 2017).

As shown in Table 3.4, for the 8-res loops, our method has a better average local RMSD compared to NGK and GalaxyLoopPS2. Though our RMSD is 0.5Å higher than the RMSD of the deep learning method Exp_GAN, we have a much smaller standard deviation as 0.21 compared to the 0.64 standard deviation of Exp_GAN.

For the results of 12-res loops as shown in Table 3.5, all four methods have similar average local RMSD around 2.2Å, but our method has the lowest standard deviation as 0.31, which is much lower than the second lowest standard deviation from Exp_GAN, which is 0.88.

Note that the Exp_GAN model was trained on a loop set which was previously filtered by the sequence alignment of the target loop. We didn't collect a training pool of such loops for each specific target loop and didn't do the sequence alignment. Sequence alignment and collecting such a training pool for each loop target were time consuming.

To fill up the missing loop gap in a protein model, we need to connect the two ends of the predicted loop fragment to the whole protein, which was not fulfilled in the ML_PULCHRA method and it only predicts the local loop structure. In the next section, we will discuss the results of ML_PLOP, which builds a whole protein structure with the predicted loop fragment included.

PDB	Loop start	Loop length	NGK	Galaxy PS2	Exp_GAN	SeqDist
135l	84	8	3.9	4.3	1.2	1.6
1alc	34	8	1.3	1.4	0.3	1.4
1btl	50	8	0.4	0.9	0.6	1.7
1cex	73	8	2.1	1.8	1.1	2.0
1clc	313	8	0.4	0.3	0.6	1.9
1ddt	127	8	3.7	1.5	1.5	1.7
1ezm	92	8	4.3	3.8	2.2	2.0
1hfc	142	8	0.7	0.9	0.9	1.4
1iab	48	8	1.0	1.8	1.3	1.4
1ivd	413	8	2.7	2.2	1.7	1.8
1lst	101	8	1.2	1.1	1.2	1.5
1nar	192	8	1.4	1.8	2.7	1.7
1oyc	80	8	1.1	1.7	0.7	1.6
1prn	150	8	8.3	8.8	1.0	1.8
1sbp	107	8	0.9	0.8	1.0	1.4
1tml	187	8	1.1	0.6	2.1	1.6
2cmd	270	8	1.9	2.3	0.9	1.9
2exo	262	8	1.5	1.1	0.6	1.6
5p21	45	8	1.7	1.9	0.5	1.8
AVG.			2.1	2.1	1.2	1.7
STD.DEV			1.9	1.9	0.64	0.21

Table 3.4: Local RMSD results of the 8-res loop set. The data of other methods was taken from Li *et al.* (Li et al. 2017)

PDB	Loop start	Loop length	NGK	Galaxy PS2	Exp_GAN	SeqDist
1a8d	155	12	3.7	3.1	2.7	2.3
1arb	182	12	1.7	1.9	2.1	2.5
1bhe	121	12	1.7	3.5	3.2	2.1
1bn8	298	12	1.1	1.1	1.7	2.3
1cb0	33	12	0.9	0.9	1.7	1.6
1cnv	188	12	6.3	6.5	2.0	2.5
1cs6	145	12	1.1	1.6	4.0	2.0
1dqz	209	12	7.5	3.3	2.6	3.1
1exm	291	12	1.1	1.3	0.6	1.9
1f46	64	12	2.6	3.8	2.7	2.5
1i7p	63	12	1.9	1.7	1.3	2.0
1m3s	68	12	3.2	2.7	2.0	2.0
1ms9	529	12	1.8	1.8	1.4	2.1
1my7	254	12	0.9	1.0	2.2	2.3
1oth	69	12	0.8	0.9	2.2	2.0
1oyc	203	12	0.7	1.2	0.4	2.2
1qlw	31	12	6.0	2.9	2.9	2.2
1t1d	127	12	1.3	1.5	2.6	2.0
2pia	30	12	0.7	0.7	1.4	2.2
AVG.			2.4	2.2	2.1	2.2
STD.DEV			2.1	1.4	0.88	0.31

Table 3.5: Local RMSD results of the 12-res loop set. The data of other methods was taken from Li *et al.* (Li et al. 2017)

3.5 ML_PLOP Results

Loop Length	Best Model Layer	Sample Count	Distance Test RMSD
5	3	19277	0.6
6	5	12853	0.7
7	5	9872	0.7
8	5	8778	0.8
9	6	7539	0.8
10	6	6330	0.8
11	8	5136	0.8
12	8	4547	0.8
13	8	3679	0.9
14	8	3223	0.9
15	9	2792	0.9
16	9	2335	0.9
17	10	1932	1.0

Table 3.6: The number of layers at the best models of each length and the corresponding test RMSD for ML_PLOP method.

The encoder block and the decoder block have the same number of layers. Table 3.6 shows the best layer number for each model at each length and the corresponding RMSD of the predicted C_α distances from the test set. The loops in the test set were removed from the training set during the sample preparation.

We then reconstructed loops by using PLOP with the predicted C_α distances as constraints on the standard 8 and 12 residues loop sets. Results are shown in Table 3.7 and Table 3.8. In the 8 residues dataset we skipped the 2sga case, because the residue length before the loop is less than 8. Our method did the side chain optimization during the PLOP reconstruction process, and we compared our results to other methods under the same scenario. Modeling

RMSDs of methods HLP-SS, NGK and Galaxy PS2 were taken from Table S4 and S5 by Park, *et al.* (Park et al. 2014). RMSDs of the lowest Rosetta-energy loop by RCD+ method were taken from Table S1 and S2 by López-Blanco, *et al.* (López-Blanco et al. 2016). All the RMSDs in the table were from the backbone C, C_α, N, O atoms.

Two rounds of PLOP were implemented: Round 1 minimized all the loop candidates generated by the loop closure stage; Round 2 applied an expensive side chain sampling to the top 20 loops predicted by Round 1 regarding their all-atom energies. The loop with the lowest energy of the two rounds was selected as the final best loop model. The average and standard deviation of all predictions were calculated.

PDB	Start	End	Modeling RMSD (Å)						
			HLP-SS	NGK	Galaxy PS2	RCD+	ML_PLOP		
							Round 1	Round 2	Overall
135l	84	91	2.5	0.3	0.4	0.5	0.7	0.4	0.4
1alc	34	41	6.9	0.3	3.1	4.6	0.2	0.2	0.2
1btl	50	57	0.9	0.4	2.4	0.6	0.5	0.5	0.5
1cex	73	80	2.5	0.3	1.1	0.9	0.5	0.4	0.4
1clc	313	320	0.4	0.4	1.3	0.4	0.1	0.1	0.1
1ddt	127	134	1.1	1.0	1.2	0.4	0.3	0.4	0.4
1ezm	92	99	0.5	0.3	2.5	0.4	0.3	0.5	0.5
1hfc	142	149	0.3	0.5	0.7	0.5	0.4	0.3	0.3
1iab	48	55	0.5	0.5	0.7	0.5	0.2	0.6	0.6
1ivd	413	420	1.3	0.8	3.9	1.7	0.9	1.2	1.2
1lst	101	108	0.7	0.5	0.6	0.6	0.3	0.3	0.3
1nar	192	199	1.2	1.3	1.1	1.0	1.2	0.8	0.8
1oyc	80	87	0.6	0.3	0.4	0.4	0.2	0.2	0.2
1prn	150	157	2.3	0.3	0.9	0.3	0.3	0.3	0.3
1sbp	107	114	0.3	0.3	0.4	0.4	0.2	0.2	0.2
1tml	187	194	1.5	0.5	1.8	0.5	0.9	1.5	0.9
2cmd	270	277	0.4	0.4	0.9	0.5	0.5	0.5	0.5
2exo	262	269	0.7	0.3	0.5	1.1	0.3	0.3	0.3
5p21	45	52	0.8	0.3	1.1	0.4	0.4	0.5	0.5
Average			1.3	0.5	1.3	0.8	0.4	0.5	0.5
Std. dev.			1.5	0.3	1.0	1.0	0.3	0.4	0.3

Table 3.7: Loop prediction performance of ML_PLOP for the 8 residues loop benchmark. Modeling RMSDs of methods HLP-SS, NGK and Galaxy PS2 were obtained from Table S4 and S5 by Park, *et al.* (Park et al. 2014). RMSDs of the lowest Rosetta-energy loop by RCD+ method were taken from Table S1 and S2 by López-Blanco, et al. (López-Blanco et al. 2016)

PDB	Start	End	Modeling RMSD (Å)							
								ML_PLOP		
			HLP-SS	Rosetta KIC	NGK	Galaxy PS2	RCD+	Round 1	Round 2	Overall
1a8d	155	166	2.8	0.6	5.2	0.5	5.2	0.8	0.7	0.7
1arb	182	193	2.6	1.4	0.4	1.0	1.4	1.5	1.6	1.6
1bhe	121	132	0.7	0.7	0.4	2.0	0.8	0.7	0.6	0.6
1bn8	298	309	2.6	0.6	1.1	1.6	0.5	0.6	0.8	0.8
1c5e	82	93	1.7	0.4	0.4	0.5	0.6	2.7	0.4	0.4
1cb0	33	44	0.3	0.7	0.6	0.5	0.5	0.3	0.3	0.3
1cnv	188	199	3.3	2.1	2.0	3.2	3.8	1.4	1.8	1.2
1cs6	145	156	3.5	3.0	2.5	3.1	4.2	1.6	1.2	1.3
1dqz	209	220	0.6	2.6	0.6	1.0	1.0	2.5	1.3	0.5
1exm	291	302	0.5	0.9	1.0	1.2	0.6	0.5	0.5	1.8
1f46	64	75	1.1	2.3	2.1	1.5	0.6	1.7	1.8	0.5
1i7p	63	74	0.3	0.4	0.4	0.3	0.4	0.4	0.5	0.5
1m3s	68	79	5.6	5.6	6.4	5.9	0.5	4.3	4.4	0.6
1ms9	529	540	2.5	1.0	2.7	2.6	0.3	1.2	1.4	1.2
1my7	254	265	0.9	2.3	0.6	2.9	0.6	2.7	0.8	0.8
1oth	69	80	0.7	0.6	0.4	2.3	1.5	0.6	0.5	4.4
1oyc	203	214	1.2	3.9	0.4	1.4	4.6	0.7	0.6	1.4
1qlw	31	42	1.4	0.9	4.8	4.6	0.7	1.1	1.2	0.8
1t1d	127	138	1.0	0.8	0.7	1.0	0.4	0.7	0.9	0.9
2pia	30	41	0.5	0.9	0.8	5.5	0.6	1.7	1.1	1.1
Average			1.7	1.6	1.7	2.1	1.4	1.4	1.1	1.1
Std. dev.			1.4	1.4	1.8	1.7	1.6	1.0	0.9	0.9

Table 3.8: Loop prediction performance of ML_PLOP for the 12 residues loop benchmark. Modeling RMSDs of methods HLP-SS, NGK and Galaxy PS2 were obtained from Table S4 and S5 by Park, *et al.* (Park et al. 2014). RMSDs of the lowest Rosetta-energy loop by RCD+ method were taken from Table S1 and S2 by López-Blanco, et al. (López-Blanco et al. 2016)

Conclusion

We showed two examples which utilized machine learning techniques on proteins.

In the first part, we designed and implemented a sequence-based method for protein-protein interaction prediction with machine learning. We have tested our methods on two datasets, the *S. cerevisiae* dataset and the human dataset, and compared the performance to other state-of-art methods. Our results are comparable to or better than other methods, according to the metrics and ROC curves.

Our research has shown that the embeddings learned from protein sequences alone with self-supervised language models can replace the MSA usage as a feature extraction method in the protein-protein interaction task. It is faster to produce the protein embeddings than the alignment method, and has similar or better prediction ability compared to using other features including MSA and empirical parameters. When the protein connection graph is dense, the embeddings from graph learning can help to increase the performance. In the future with the development of the embedding models and the increase of the training samples, our methods should achieve better results.

In the second part, we proposed and implemented an accurate and fast loop modeling method combining machine learning and PLOP.

A complete and accurate protein model is useful for computational biologists to simulate the protein physiological states and functioning process. We developed an accurate and fast prediction method for inter-residue distances within loops. The input is the primary sequence of the loop and its flanks. For the interval length longer than 2 residues, the predictions have the correlation coefficients higher than 0.9 for the models of loop length between 5 to 17 residues. The local RMSDs of the reconstructed loops by MDS and PULCHRA from our method are competitive to other methods, and our standard deviations are the lowest among all the methods. The downside of this method is the difficulty to connect the two ends of the loop back to the whole protein structure. We then integrated our method as distance constraints to accelerate the *de novo* loop modeling method PLOP with the help of Dr. Tianchuan Xu at Schrodinger, Inc.

We integrated the modern machine learning techniques into PLOP to address the difficulty of loop prediction. With our protocol, we not only make use of the protein loop template information but also retain the flexibility and accuracy of *ab initio* method at a lower cost. For the 8 and 12 residues loop sets, our ML_PLOP method achieved RMSDs of 0.5 and 1.1 Å with standard deviations of 0.3 and 0.9 Å. In the 8 residues loop set experiment, our result is one of the best and similar to NGK method. In the 12 residues loop set experiment, our method clearly beats other methods on both the accuracy and deviation. Note here NGK method takes a long time to run, and our method has a lower cost after being accelerated by the machine learning algorithm. RCD+ is the secondly best method in the prediction of the 12-residue loop set, but it is limited to the loops with length lower than 12 residues.

www.ingramcontent.com/pod-product-compliance
Lightning Source LLC
LaVergne TN
LVHW041750190726
843493LV00008B/2539